The Buffalo Bill Historical Center acknowledges IRMA LAROM for her generous and enthusiastic support for this project.

I.H. "Larry" Larom and Irma Larom in front of their Valley Ranch home, c. 1930.

WELCOME TO MY WEST

I. H. LAROM:

DUDE RANCHER, CONSERVATIONIST, COLLECTOR

TEXT BY
Lawrence R. Borne

BUFFALO BILL HISTORICAL CENTER, CODY WYOMING

Manufactured in the United States of America

Library of Congress Catalogue Card Number: 82-71998
I.S.B.N. 0-931618-07-X

(Cover) **Beaded Crow belt pouch,** c. 1930. **(Photo by Mark Payler)**

CONTENTS:

Young visitors at Vally Ranch, c. 1940.

The Board of Trustees

Preface

Larry Larom once told a Miami journalist that he might never have made it to Wyoming had it not been for the Grand Duke Michael of Russia. While on a trip to Egypt and the French Riviera in 1912, Larom and his father were persuaded by the Grand Duke to postpone their return to the United States. The ship they were to sail on was the Titanic.

We owe a debt to that Russian nobleman. Larry Larom did travel west, and he made his home here for over fifty years. *Welcome to my West* is an invitation to visit and savor the West of Larry Larom. It is a West which he saw change dramatically during his lifetime, and whose heritage, color, and hospitality he helped preserve.

Western writer, J. Frank Dobie, remarked in the Fort Worth *Star Telegram* that he saw more antelope than people between Cody and Valley Ranch. The South Fork of the Shoshone River still teems with wildlife. Bounded by towering mountains, the green, lush river valley is a treat to all the senses. Valley Ranch is located near the end of the sixty mile stretch and has remained the same oasis of solitude and beauty that captivated Larry Larom in 1911. He and his partner, Winthrop Brooks, could not keep Valley a secret for long after they bought it. The songs of the meadowlarks and the bugle of the elk were soon mixed with the shouts and laughter of eastern dudes trying their hand at fly fishing for native trout, riding horseback on the trails that snake through the mountains and alpine meadows, and helping the wranglers brand cattle. These were unusual pastimes for folks like Jack Bouvier, whose young daughter Jacqueline went on one of Valley's famous three week pack trips; or for Lucius Boomer, president of the Waldorf Astoria where the Laroms maintained a suite for the New York visits.

Larry's father had advised him as he launched his dude ranching career, "Make your guests measure up to your standards." If the Guest Book at Valley Ranch reads like a social register, it is because most of the guests were actually listed in the New York *Social Register*. In his own home at the ranch, Larry never sacrificed the atmosphere of formality in which he had been reared. Anthony Gerharz, who with his father was Larom's accountant, relates that he was once granted special dispensation when auditing the books at Valley Ranch. Larry called him at the ranch office to announce cocktails and supper. A few minutes later he called back and added, "You needn't dress for dinner."

Larry realized that some measure of formality helped his eastern guests make a more comfortable transition to the easy-going ways of a western ranch vacation. He structured their activities so that should they prefer not to, they might never have to make decisions about how to spend their days. He prided himself on attending to details, even arranging the seating in the dining hall, in order to ensure a care-free ambience. His efforts worked. Entire families made annual pilgrimages to Valley. Sons and daughters grew up and returned with their own families to Valley. Mrs. Anne Kneeland Ellis wrote Larry, "We are looking forward to the time when we can bring our child (or children, by that time) to the Valley Ranch so that they might have as much fun as we had!"

Valley Ranch was a cosmopolitan center in the midst of the wilds of Wyoming. Nearby ranchers often brought their guests to Valley for an evening's entertainment. Actress Hope Williams Read, a neighbor of the Laroms, introduced her circle, including Noel Coward, Clifton Webb, and Tallulah Bankhead, to the New Yorkers and titled Europeans at Valley. They danced to hit tunes played by a band which might be led on any given night by a musician from the New York Philharmonic. Or they listened to a wrangler strumming his guitar and singing cowboy tunes alongside a campfire while the Wyoming night sky blazed back with a thousand stars. Valley Ranch was not just a place but also an experience which created generations of memories for the dudes.

Larry Larom is remembered for much more than being the host of one of Wyoming's earliest and best known dude ranches. He helped found the Dude Ranchers Association in 1926 and guided it through its most difficult years. He ensured that the industry would be represented by ranchers whose standards of hospitality were as high and uncompromising as his own.

Larry's love for the outdoors translated into a deep and genuine concern for the conservation of land and wildlife. He had been a city boy. An uncle introduced him to the joys of hiking and camping and instilled in him an appreciation of nature. As a young teenager, Larry organized long backpacking trips through the Adirondacks. As an adult and a westerner, his special knowledge and feeling for the oudoors drove him to work to preserve the wilderness for succeeding generations.

He was a keen observer of western peoples and feared for the loss of their material culture. He began collecting Indian artwork and artifacts upon his first visit to Wyoming. Sincerely interested in saving reminders of passing ways of life, he amassed one of the finest collections of Indian and cowboy materials in the West. His trading parties with nearby Indian tribes, described in the following essay, were among the most colorful episodes in Valley's history.

Cody and Northwestern Wyoming were also beneficiaries of Larry's devotion to his adopted home. He was a co-founder of the Shoshone River Power Company which brought electricity to residents of the South Fork. He was a vital force behind the incorporation of the Cody Stampede. His efforts to improve the airport helped persuade Frontier Airlines to inaugurate regular passenger service to Cody. His unfailing support of the Buffalo Bill Historical Center contributed to its becoming one of the nation's oustanding museums.

All of us who love the West—its scenery, its wildlife, its history and color—owe a debt to Larry Larom, an eastern dude who became a western institution.

Welcome to My West is a tribute to a great westerner. It is with special thanks we recognize Mrs. Irma Larom for her help in organizing and promoting the exhibition. Thanks also go to the staff of the Buffalo Bill Historical Center, Dr. Larry Borne for writing the text, and Mark Payler for the photographic work. And finally, we sincerely thank all the friends of Larry Larom for giving us the opportunity to learn more about his personality and accomplishments.

Michael Kelly and Paul Fees
Buffalo Bill Historical Center

Valley Ranch and the South Fork of the Shoshone River, 1950.

I. H. LAROM:

DUDE RANCHER, CONSERVATIONIST, COLLECTOR

by Lawrence R. Borne

"Am leaving for a Game and Fish Commission meeting in Cheyenne on the 12th of November and returning on the 16th, then to Dude Ranchers Association Convention in Sheridan on November 20th to November 25th."

Larry Larom
November 10th, 1957

This sentence summarizes much of the life of one of the most prominent residents of the Cody, Wyoming region for six decades of the 20th century. Irving H. (Larry) Larom was a westerner by choice rather than by birth, but he adopted his new home enthusiastically and became one of its most ardent boosters.

Larry Larom, c. 1930.

Larry Larom and his sister, Edith, c. 1895.

Larry Larom and friends camping in the Adirondack Mountains, New York, c. 1913.

The letter was addressed to Harold McCracken, an authority on Western art who would soon become Director of the Whitney Museum of Western Art, and it illustrates three important facets of Larom's western life. Larom had been a friend of Colonel William F. Cody (Buffalo Bill), and as a collector and donor of Western Americana and Indian artifacts, became a life-long friend and benefactor of the Buffalo Bill Historical Center. Larom's reference to the Game and Fish Commission meeting in Cheyenne is indicative of his constant interest in conservation activities of all types. And the reference to the Dude Ranchers Association (DRA) Convention in Sheridan points to the center of his business life in the West: dude ranching. Larry Larom was not only a dude rancher, he was also one of the chief promoters of this burgeoning industry in the early 20th century and became its principal spokesman. He was one of the founders of the DRA in 1926 and served as its president through 1944.

Irving H. Larom was born in New York June 3, 1889 and was educated at Lawrenceville School and Princeton University. From his youth he was a lover of the outdoors and he used to organize hiking trips of ten to twenty days through Vermont, New Hampshire, and Maine as well as in the Adirondack and Catskill mountains. This obviously prepared him for an outdoor life but he was drawn west by two other influences.

One of these was a print of an Albert Bierstadt painting showing the scenic beauty of the American West which hung in Larom's bedroom in his New York home. The second influence was far more lively. Larom attended several shows of "Buffalo Bill's Wild West and Congress of Rough Riders of the World" featuring not only cowboy fun, trick shooting, and Indian attacks, but the great frontiersman himself, Colonel William F. Cody. A few years after his first glimpse of the show, Larom was taken behind the scenes and introduced to Cody and other members of the cast; he was even invited to ride in the Deadwood stagecoach. The action and excitement certainly piqued the young man's interest, but even at this show a painting enhanced the appeal of the West. A painted canvas backdrop the length of a city block was used in the show to portray the mountains along the South Fork of the Shoshone River near Colonel Cody's TE ranch. Young Larom had met Buffalo Bill, Annie Oakley, and Johnny Baker, but this painting helped convince him that it was time to see the West.

Above:
Indian Encampment in the Rockies,
by Albert Bierstadt. Steel engraving,
15¾ x 27½ inches, 1866,
Private collection, Cody, Wyoming.
(Photo by Robert Weiglein)

Left:
Buffalo Bill and Indians
at the Wild West, c. 1910.

Winthrop Brooks and Larry Larom at Valley Ranch, c. 1915.

In 1910, while a student at Princeton, Larom headed for Wyoming. His first western experience must have been a bit unnerving since he was simply dropped off the Burlington Railway train at 3:00 a.m. near a water tank. But he took his unorthodox arrival in good humor, enjoyed a breakfast of chicory coffee and apple pie with a Swedish track gang, and then boarded a local train for Cody. This northwestern Wyoming town was a lively place with Buffalo Bill himself the genial host at the beautiful Irma Hotel. Larom obviously enjoyed his first visit there since he returned in the summers of 1911 and 1913. In 1914 he brought a friend, Winthrop Brooks, with him. The young men stayed with Jim and Jenny McLaughlin who lived in a sod-roofed cabin on their homestead in the spectacular valley of the South Fork of the Shoshone River. Brooks was also impressed with Wyoming, and in 1915 he and Larom acquired the McLaughlin homestead: two small cabins and 74½ acres of land. Brooks and Larom began raising cattle on their place, which they called Valley Ranch. However, they were soon entertaining friends from the East who wanted to sample rustic life on a Wyoming ranch, so the two men made Valley a dude ranch as well.

The word "dude" and the phrase "dude ranch" conjure up some disaparaging images among people in the latter part of the 20th century; but in 1915 the word "dude" just meant

Valley Ranch soon after its purchase by Brooks and Larom, c. 1915.

someone from the East who paid a westerner for food, lodging, use of a horse, and/or guiding services. Larom and Brooks had been dudes on their first western visits. A dude ranch was simply one that accommodated guests. The Valley Ranch's quick adaptation from cattle operation to a cattle and dude ranch was much like that of other stock ranches that had begun accepting paying guests during the three previous decades. These guests gave ranchers an extra source of income and even saved some of them from bankruptcy when the cattle industry suffered serious declines.

Larom fit into the western atmosphere easily and was soon known by the nickname "Larry." He spent more time in Wyoming than Brooks and took direct charge of the ranch operation. Life on the Valley Ranch was pleasant, and the number of visitors rose so much from 1915 to 1920 that guests became the most important part of the ranch's business. This increase stemmed largely from the effects of World War I. Travel to Europe was virtually halted and numerous easterners turned west to find suitable vacation spots for themselves and their families.

Valley Ranch's first guests, c. 1915.

Larry Larom, 1910.

Irma Larom,
drawing by Norma Virginia Sweringen
Charcoal on Paper, 24" x 16", 1936
Courtesy of Mrs. Irma Larom
(Photo by Robert Weiglein)

Valley Ranch "Stage," c. 1920.

The war took Larom away from the West for a while since he joined the Army in 1918, but Win Brooks kept the ranch operating efficiently. Larom returned to Wyoming after the war and married a young Cody woman, Irma Elizabeth Dew; they were to make the Valley Ranch their home for more than fifty years. Larom and Brooks needed more land to accommodate the growing number of guests and to raise more cattle and horses. So they acquired adjoining land and constructed cabins and other necessary buildings. By 1927 they owned eighty-three buildings and owned or leased 8,000 acres of land.

While the partners appreciated the beauty of their surroundings, they learned the great difficulty of making a living on a western ranch. Valley was forty-five miles of dirt road from town where they got supplies and picked up guests for the ranch. Travel was complicated by the long winters, spring floods, and summer cloudbursts. The romance of ranch life may have worn thin for Larry Larom but it was replaced by a tough determination to succeed and a realistic view of what was needed for success. Dude ranching was primarily a summer operation; families would come for weeks or even months at a time when their children were out of school and when Wyoming weather was generally warm and pleasant.

The Valley Ranch entertained many families but also launched special programs for teenagers. In 1922 Larom and Brooks developed horseback pack trips through Yellowstone Park and the Jackson Hole country. There were two separate trips each summer, one for boys and one for girls, each averaging sixty youngsters. A staff of counsellors, a crew of eighteen men, thirty team horses to pull the cook and supply wagons, and over one hundred saddle horses were required for these trips. Eventually, these youth trips would last forty-five days, giving the boys and girls a lengthy experience in wilderness living. Other pack trips were also arranged through the Shoshone and Teton National Forests. These, along with the daily horseback rides for people who stayed on the ranch, required at least one hundred additional horses. Thus, the Valley Ranch had become a beehive of activity less than a dozen years after its founding.

As the ranch was developing, the owners received requests to take boys during the fall and winter months so they might benefit from a longer contact with ranch life. These inquiries prompted Larom and Brooks in 1922 to open a unique college preparatory school. Larom had graduated from Princeton and Brooks from Yale; Julian S. Bryan a friend associated with them in this venture, had graduated from Colgate and had served as the principal of Pelham Manor School. The three men asked their contacts at several important eastern schools to recommend boys for the school at the ranch. Larom also recruited among the summer families at Valley and from the boys who went on the pack trips.

Dude preparing for a three week pack trip. c. 1945.

Dudes in Grand Teton National Park, c. 1945.

Valley Ranch School students in Brooks House classroom, c. 1930.

Valley Ranch School polo team, c. 1930.

The school developed a four-year college preparatory curriculum modeled after eastern schools and geared toward the College Entrance Examination Boards. It opened October 1, 1922 with five students. Enrollment doubled the second year and eventually rose as high as thirty boys taught by a faculty of six men. As headmaster and director, Larom hired staff from Harvard, Princeton, Penn State, and the University of Illinois to teach English, history, mathematics, physics, Latin, German and French. The Valley Ranch School combined academic work and discipline, and outdoor activities were required for the boys each afternoon. These sessions included shooting on the rifle range, horseback riding, trapping, fishing, mountain climbing, football, soccer, baseball, and basketball. Polo was later added as well. In addition to these regular activities, there were occasional all-day horseback rides and weekend pack trips.

To meet the needs of the school, Larom constructed or modified buildings, improved plumbing facilities, and installed electricity — expensive undertakings in such a remote location. There were also salaries to pay to both the masters who taught the young men and to the full-time nurse who looked after their health. Each student was charged $1,550; this included tuition, housing, food, books, supplies, and the use of a horse and saddle. Students furnished their own clothing and a .22 caliber rifle. The Valley Ranch School was so successful that by 1925 it was one of only three institutions in the Rocky Mountain area where the College Entrance Examination Boards were given. Soon its graduates were matriculating at Princeton, Yale, Harvard, Dartmouth, Brown, and other colleges. The school operated during the fall, winter, and spring months except for a four weeks' vacation at Christmas. As it grew through the 1920s, it was moved to the beautiful, spacious home that had been built for Winthrop Brooks when he stayed on the ranch.

Brooks was very important in the acquisition and early operation of the ranch, but he had to leave it to assume the presidency of the famous Brooks Brothers clothing store in New York City. Larry Larom purchased his partner's interest in 1926 and became sole operator of Valley Ranch. Brooks remained a good friend, and Brooks Brothers provided the Valley Ranch with office space in New York City for many years. This gave Larom a valuable, centralized location in the nation's largest city.

Even as these notable changes were occurring, Larry Larom was becoming a most vigorous promoter of dude ranch vacations in general. Since he was from the East and had himself been a dude, he clearly understood what Easterners thought about the West and what it would take to encourage more of them to try a ranch vacation. He realized he and other ranchers had to nurture the West's fascination for people and also to dispel the wild and woolly image that had been prevalent in the East.

Personal contact was the key ingredient in successful dude ranch advertising. Dude ranches had developed through friendships and hospitality, so it was natural that they continue to grow that way. Larry Larom had probably the best eastern arrangement possible because of his year round office in New York City. He kept a secretary there to answer inquiries and to handle problems that past and prospective guests might have. Larom himself spent several weeks in New York each spring to take a direct hand in getting and keeping ranch business, valuable for the personal contact that it permitted. Dude ranchers wanted to feel they were inviting friends into their homes. Ranch vacations tended to be rather extensive. It was common for families or individuals

Brooks Brothers Trademark

Evening entertainment at Valley Ranch, c. 1945.

Larry Larom promoting Wyoming during a visit with President Calvin Coolidge, the White House, Washington, D.C., 1927.

Northern Pacific Railroad advertisement for Valley Ranch, c. 1931.

to remain from three weeks to an entire summer on a dude ranch. Obviously such a lengthy stay could become quite trying if there were significant personality clashes or if there were misunderstandings about what ranch life was like.

However, Larom did more than seek business for the Valley Ranch while in the East. He helped all dude ranches by getting information published about this growing industry. He was interviewed by various reporters, and the New York newspapers ran features with attractive photographs of dude ranch life. While newspaper stories were useful, Larom and other dude ranch promoters found that magazine stories or advertisements were even more valuable, particularly in magazines that catered to the type of people who had the means to take a long western vacation.

Larry Larom was fortunate in beginning Valley Ranch when he did. The vigorous spirit of the 1920s encouraged outdoor activities and family vacations and, of course, there was a general feeling of optimism and prosperity. Thus, although there were occasional slumps in the economy, the 1920s were the golden age of dude ranching. Dude ranches were started in many western states but they were most numerous during this decade in Wyoming and Montana.

While Larry Larom was not the first in the field, he became the most prominent spokesman and promoter for the industry for many years. Many new ranches were begun during the decade, and Larom and others recognized in 1926 that mutual problems made the time ripe to organize this growing industry. The two ranchers who seemed to be the most instrumental were Larom and Ernest Miller of the Elkhorn Ranch near Bozeman, Montana. Railroads also were interested in a dude ranch organization. While many railroads eventually became involved, the most important one at the beginning was the Northern Pacific. Max Goodsill, the

General Passenger Agent, and A.B. Smith, the Passenger Traffic Manager, helped organize the meeting of ranchers and railroad officials in Bozeman, Montana in September, 1926. The Northern Pacific was beginning to feel the competition from automobile travel, so its promotion was quite logical an activity to help offset this competitor.

Dozens of people met in Bozeman that autumn to become better acquainted and to discuss how they could more effectively develop dude ranching. The ranchers decided to call their new organization the Dude Ranchers Association (DRA) and emphasized several key aims: classification of ranches; active support for the protection of game and fish; cooperation with the Northern Pacific and other organizations to promote travel to ranches; advertising, and personal solicitation of prospective guests. The ranchers chose Larry Larom as their first president, A.H. Croonquist of Camp Senia, Montana as vice president, and Ernest Miller as secretary-treasurer. Thus the first dude ranch organization was launched. Larry Larom would remain its president continuously through 1944, and he became one of the prime shapers of this new group. His title was not an honorary one since he was a vigorous worker and enthusiastic organizer. The DRA got a big boost in 1928 when Larom and Max Goodsill persuaded T. Joe Cahill to become the group's executive secretary, a position that would require a great deal of travel and hard work. The Northern Pacific proved its continuing support by giving Cahill inter-railroad passes and paying part of his expenses. In 1929 the DRA began publishing Minute Books of its meetings. Three years later, the organization started its beautiful and informative magazine, *The Dude Rancher*. The Minute Books and magazine included data about the economic importance of dude ranches plus reports on committees and legislation of interest to ranchers.

The Dude Rancher magazine, Vol. 1 #2, March 1933 (Photo by Robert Weiglein)

Vegetable gardens at Valley Ranch, c. 1940.

Making hay at Valley Ranch, c. 1930.

Under Larom's leadership the DRA was immediately successful and membership rose to ninety-one ranches by the end of the 1929 season. It was fortunate that the DRA began so well, for dude ranching now faced an economic challenge at the end of the decade. There was no catastrophic decline in dude ranch guests in 1930 but, as economic woes grew into the Great Depression, dude ranches suffered just as other businesses did. Larry Larom remained at the head of the DRA throughout the troubled thirties and, while the industry and the organization survived, some ranches did not. One that did was Valley Ranch.

Valley Ranch had become a nearly self-sufficient operation over the years with eighty dairy cattle and two-acre vegetable garden, in addition to other facilities, to feed its many guests. The Valley School, however, became a victim of the Depression. There just were not enough families who could afford to send their sons west during these harsh years and the school ceased operation in 1934. The Depression also brought an end to a unique service the Laroms had provided to bring guests to their ranch. They had arranged for special Pullman cars and a diner to be attached to a train in New York City and pulled all the way to Cody, where the guests were met and driven to the ranch. Declining numbers of guests in the 1930s made this special service impractical. The Laroms had faced tough times before, and they were able to survive these hardships. Then, just as business seemed to be returning to normal at the end of the 1930s, the nation and dude ranching were hit again — this time by World War II.

The Second World War brought numerous problems to dude ranching. One of the most obvious was the lack of ranch workers. So many young men entered the armed forces that ranchers had to use men too old for the draft or do more of the work themselves. Another major problem ranchers faced was the proliferation of government bureaus and regulatory agencies. Rationing of automobile tires and gasoline and other limitations on travel were particularly damaging to dude ranches. Wartime restrictions made it difficult for ranchers to get supplies even when they had the opportunity to do so. Larry Larom who was still the president of the DRA, urged his fellow ranchers to gather data on their food production. He wanted to be able to show that dude ranchers were important to the war effort and that they deserved some relief from government-imposed travel sanctions. Many dude ranches, including Valley, were fortunate that their cattle operations thrived as prices for beef rose with war-time demands. The DRA magazine, *The Dude Rancher*, featured a two-page spread titled "The Dude Ranch in War Time." This article highlighted the fact that dude ranches supplied beef and other food for the country and also offered recreation for soldiers and respite for other citizens from the tensions of the time. President Franklin Roosevelt had stated it was proper for people to engage in recreational activity when possible, and ranch advertising stressed that guests should not feel guilty about going to ranches when they had the opportunity. They were asked to realize, however, that workers were difficult to find and guests could not expect extra service at ranches. Railroad officials strove to accommodate vacation travelers as wartime obligations allowed. Of course, Larry Larom as president of the DRA was in the midst of many of the discussions about wartime travel and problems.

At his own Valley Ranch, Larom altered his operation to jibe with the war effort and in the early 1940s published a special brochure, "Summer Suggestions." In it he described the restrictions and limitations, but he assured the reader that

Summer Suggestions, c. 1943.

Deer at Valley Ranch, c. 1950.

vacations were still available. He also promised that pack trips and big game hunts could be arranged when people made early reservations as long as wranglers and guides could be hired. One of his innovations was the Victory Vacation. Inaugurated in 1943, it offered a six or eight week visit to the ranch for boys and girls. Prices for the stay were lower than normal but the youngsters had to care for their own cabins and devote several hours per day to work in the gardens, hay fields, and corrals. This unique program demonstrates the versatility and adaptability of Larom and other dude ranchers in surviving wartime changes.

Survive they did, and they faced a new America. After the great crusade of World War II was over, Americans eagerly abandoned the travel restrictions and vacationed throughout the country. Dude ranches were ready to cater to these vacationers, and many ranches regained much of their pre-war business. Ranches in Montana and Wyoming did not benefit as much as did those farther south. New dude ranches were developed, especially in Colorado, Arizona, and New Mexico, and many people turned to those areas where they or their relatives had been stationed during the war years.

Wooden gate sign. Valley's "Lazy JD" brand originally belonged to South Fork homesteader John "Reckless" Davis.

Larry Larom stepped down from his position as president of the DRA near the end of the war, although he always remained involved with, and interested in, its operation. He and his wife, Irma, continued to operate the Valley Ranch which was still their home.

Even as he continued his dude ranching, Larry Larom also remained active in his other realms of interest. One of these was conservation. To Larom and other ranchers conservation meant preservation of the land, controlled hunting and fishing, and protection of wild animals and fish.

Of course, conservation activities by dude ranchers contained an element of self-interest. Adequate game not only assured successful hunting and fishing trips but also attracted many guests who just enjoyed seeing the wild animals in their natural habitat. But ranchers were motivated by more than the expectation of economic benefits. Love for the land and animals had been a factor in drawing the ranchers to ranch life, so it was quite natural that they try to preserve both, however they could. In his notable book, *American Sportsmen and the Origins of Conservation*, John F. Reiger has made a strong case for sportsmen being the orginators of the conservation movement in the United States. Dude ranchers and their guests clearly fit into this story and reinforce Reiger's thesis.

Among dude ranchers some of the most notable conservationists were Howard Eaton, Dick Randall, Charles Moore, Struthers Burt, and Larry Larom. Like most dude ranchers Larom was a transplanted easterner, but his interest in conservation was just as intense as that of natives like Charles Moore. As early as 1916, soon after starting the Valley Ranch, Larom had expressed concern about the preservation of big game animals. This interest was undoubtedly reinforced by

Larry Larom with trophy bull elk, c. 1925.

Larry Larom and conservation officials, c. 1930. (L to R) T. Joe Cahill, Secretary of the Dude Ranchers Association; Ray Lyman Wilbur, Secretary of the Interior; Larry Larom; H.M. Albright, Director of the National Park Service.

his friendship with William Cody, and it ultimately became interwoven with his dude ranch work. When Larom went East to publicize dude ranching and sign up guests, he also set aside time to visit government officials to discuss matters relating to national parks. Through his Valley Ranch School, Larom was also a teacher about outdoor life and the proper and safe use of the land.

The Dude Ranchers Association was also extremely active in the conservation field. One of the important participants at the founding meeting of the DRA in 1926 was Horace N. Albright, Superintendent of Yellowstone National Park. Dude ranchers and their guests were probably the prime users of Yellowstone Park in the 1920s, and there was a cordial relationship between the ranchers and Albright. As noted, dude ranchers at the first DRA meeting included among their goals the protection of animals. One of four general committees created at this initial meeting was organized to influence legislation regarding game and fish. The desirability of involvement was re-emphasized in the DRA constitution written some months later. One of the organization's four stated purposes was to cooperate with officials of the national parks and forests for conservation and preservation of the land. Later, the organization granted a number of honorary memberships, and many were given to men connected with conservation: the Director of the National Park Service, supervisors of national forests, game wardens and protectors, and members of state fish and game commissions. During Larom's presidency, many of these people were prominent speakers at the DRA Convention in the fall of each year. They explained their particular functions, praised the dude ranchers' activities, and promised cooperation with them.

So interested were the ranchers in these activities that, at

Dudes fly fishing along the South Fork of the Shoshone River, c. 1935.

A day's catch at Valley Ranch, c. 1935.

Elk wintering in Jackson Hole, Wyoming, c. 1950.
(Photo by Charles Belden, Charles Belden Collection, Buffalo Bill Historical Center)

the sixth annual meeting of the DRA, President Larom announced that the first aim in their program had been met — the recognition of the dude ranch as a valuable factor in the development of the western states. Larom stated that:

> *Our present efforts are being directed towards the building up of a closer understanding of the problems in our districts, with particular reference to the restocking of streams and proper conservation of our game and fish resources.*

To support these goals the DRA sometimes openly engaged in political affairs. It passed resolutions urging Congress to appropriate specific sums of money for conservation projects. Larry Larom frequently visited Washington, D.C. to meet with state congressional delegations as well as officials of the forest and park services. In the late 1930s, the DRA opposed a Forest Service re-organization bill, fearing that the agency would no longer function as well as it had in the past. The DRA was also involved in state politics, primarily in Wyoming and Montana. Larom served on the Wyoming Fish and Game Commission and appointed other members of the DRA to the Advisory Board of the Washakie National Forest. He assisted game officials in feeding the elk herd in Yellowstone Park in the 1930s, a problem that dude ranchers had addressed before when harsh winters had limited access to the natural food supply for this huge herd. In the 1950s Larom vigorously protested an attempt by the federal government to require licenses for recreational uses of land in the national forests.

Thus, during and after Larom's presidency the DRA was clearly a major force in the conservation field. The DRA minutes show page after page of discussions on fish and game matters. The members and the organization supported hunting and fishing, but strictly within legal limits in contrast to the earlier days of unrestricted destruction of game and fish. Ranchers often acted as deputy fish and game wardens to augment the slim ranks of state and federal game officials. They also helped officials stock fish in lakes and streams, and they helped fight forest fires that threatened national forests and parks.

Larom was also active in various other conservation organizations. He was a charter member of the Izaak Walton League; he helped develop the American Forestry Association and was made an honorary Vice-President in 1945; he was appointed Chairman of the Federal Elk Commission; and he was a director of the American Wild Life Institute which held its annual meeting at the Valley Ranch in 1952. Indeed, Larry Larom was so esteemed among conservationists that in 1940 Secretary of the Interior Harold Ickes invited him to Washington and offered him the position of Director of the National Park Service. Larom was unwilling to accept this offer since he believed it would put too much strain on his family and business life. Ickes told Larom that the "Park Service would have benefitted greatly from your service . . ." and urged Larom to keep in touch and to visit him in Washington.

Larom maintained his contact with Ickes but never took a job in Washington; he was committed to the West and kept his home at his ranch. He remained dedicated to conservation, however, and he devoted as much time and effort to it as before. At the same time he continued to run his successful business, even while presiding over the Dude Ranchers Association.

"The tremendous interest the West has in the good administration of the National Forests appeals to you as it does to me, and I hope most earnestly that our joint efforts may help . . ."

Gifford Pinchot
September 21, 1939

Izaak Walton League award presented to Larry Larom for his contributions to conservation, 1965. (Photo by Robert Weiglein)

Crow Indians dancing for guests at Valley Ranch, c. 1930.

Crow Indians with travois arriving at Valley Ranch, c. 1930.

Larry Larom was not only a public-spirited businessman and conservationist. He was also an ardent collector of American Indian materials. More than a hobbyist, Larom felt an obligation to preserve the art and artifacts of a fast disappearing culture for future generations. Few people outside his immediate circle of friends and acquaintances were aware of his interest in the subject; and even fewer knew how extensive and excellent his collections were.

Larry Larom first saw Indians in New York City—as members of Buffalo Bill's Wild West. Larom later said that he started collecting Indian artifacts in 1910, as soon as he came to Wyoming. After he established the Valley Ranch, Larom had opportunities to acquire more material from the Indians. He invited Crow and other Plains Indians to the ranch, and they pitched their tepees on the lawn near his home where they would stay for a week or ten days at a time. One special occasion for their visit was always at the end of the long pack trips which were a feature of the Valley Ranch. When the riders returned, people from the surrounding area, including nearby Indian groups, would come to Valley for a rodeo and celebration. The arrival of Indians in their ceremonial costumes was one of the most striking and memorable experiences for summer guests at Valley Ranch. As he made friends among the Indians, Larom traded food, money, books, and other items for necklaces made of bear claws, knives and knife cases, headdresses of golden eagle feathers, buckskin scabbards, clothing decorated with beads, and items relating to Indian "medicine." Larom used some of the paraphernalia to decorate rooms at the Valley Ranch, but that represented only a fraction of what he had accumulated.

Larom's interest in the Indians was influenced by his friendship with Buffalo Bill Cody. Eventually the memory of his friendship and his collection of Indian artifacts would merge in the Buffalo Bill Historical Center. In January, 1917, just weeks after Buffalo Bill's death, a group of his friends, including Larom, gathered in Cody and decided that something lasting should be done in his remembrance. They formed the Buffalo Bill Memorial Association and received a charter from the state to build and maintain a monument to William F. Cody dedicated to the preservation of western history. Larom's early partner, Winthrop Brooks, was also involved in this endeavor and in 1923 was elected Chairman of the National Committee of the Buffalo Bill Memorial Association. Others on this committee were General John J. Pershing and Daniel Beard, founder of the Boy Scouts of America. The committee examined the heroic bronze statue of Buffalo Bill created by the renowned socialite-sculptor, Gertrude Vanderbilt Whitney. This equestrian state, *The Scout*, was brought to Cody and dedicated in 1924. It was placed on a forty-acre plot of ground donated by Mrs. Whitney to the Buffalo Bill Memorial Association in honor of the frontiersman.

The center to honor Colonel Cody grew slowly, but Larry Larom was a constant supporter of it. He wrote numerous letters asking people to donate or lend appropriate material to the newly-built Buffalo Bill Museum. He also encouraged his ranch guests to visit the museum. Meanwhile, he continued his own collecting, even going among the Indians of the Southwest to gather Navajo rugs, kachina figures, Kiowa belts, and miscellaneous items from the Apache and Zuni tribes. He became an authority on Indian artifacts and advised museum officials on the value of potential acquisitions. In the 1950s the Memorial Association named him a trustee of the museum and Chairman of the Relics Committee. In this capacity he became even more active in his quest for materials suitable for display in the museum.

Larry Larom (far left) and other dignitaries, at the dedication of the Buffalo Bill Museum, July 4, 1927.

A selection of Indian materials at the Valley Ranch trading post, c. 1945.

Plains Indian Museum, Buffalo Bill Historical Center, Cody, Wyoming (Photo by Robert Weiglein)

But Larom's involvement went far beyond urging other people to make donations. He lent many items from his own collections to the museum. These were especially appreciated in the 1930s and 1940s when museum holdings were quite limited. In the late 1950s Larom decided that a catalogue and appraisal should be made of his collections.

Officials from M. Knoedler and Company and the Museum of the American Indian - Heye Foundation, both of New York City, examined and appraised the material he had lent to the Buffalo Bill Museum. The resulting catalogue covered 25 pages and included 425 separate listings. These items ranged from a heart-shaped medallion to a Sioux woman's dress of blue and red flannel embellished with 350 elk's teeth. While these artifacts had previously been loaned, Larom now began donating a portion of his material to the museum each year. Gradually, all of this magnificent collection was transferred to the Buffalo Bill Memorial Association and designated "The Irving H. 'Larry' Larom Indian Collection."

The collection was displayed on a substantial scale in 1959 in the first large, modern museum building opened to the public in Cody. Constructed on the forty acres given years earlier by Mrs. Whitney, this was the Whitney Gallery of Western Art. Larom's Indian Collection generated discussion about building an adjacent Indian museum. As officials of the Memorial Association worked towards this goal, other private collections of Indian artifacts were donated and funds were given for the acquisition of Indian relics. In 1979 the Plains Indian Museum of the Buffalo Bill Historical Center was opened to the public. Even with all the material that has been gathered over the years, one of the more significant collections remains that which was donated by Mr. and Mrs. Larom.

Mrs. Larom still lives in Cody, Wyoming and spends her summers on the Valley Ranch amidst the breathtaking scenery along the South Fork of the Shoshone River. Larry Larom died at his ranch at the age of 84 in December, 1973, but much of what he built remains. The Dude Ranchers Association, which Larom helped found in 1926 and which he presided over through 1944, continues to operate. Its magazine, *The Dude Rancher*, is still published and has honored Larom many times in its pages.

In keeping with Larom's long association with the Buffalo Bill Historical Center, Mrs. Larom donated his business and personal records to the Center. Designated the "Irving Hastings (Larry) Larom Manuscript Collection," it is located in the Harold McCracken Research Library.

Larom's collection of Indian artifacts can be seen at the Plains Indian Musuem of the Buffalo Bill Historical Center. The Center proudly keeps a large bronze plaque given in Larom's honor by his many friends. Mrs. Larom has continued to donate to the Center additional Indian artifacts and other items from her husband's collections.

As dude rancher, conservationist, and collector of Indian artifacts, Larry Larom did much to preserve the western culture and landscape that had drawn him to Wyoming. Present-day visitors can enjoy what he saved and be grateful that he journeyed West to stay in 1915.

Lawrence R. Borne
Department of History
Northern Kentucky University

Larry Larom at Valley Ranch, c. 1970.

Bibliography

The bulk of the material about I.H. Larom is in the "Irving Hastings (Larry) Larom Manuscript Collection," housed in the archives of the Buffalo Bill Historical Center. This collection was organized in 1980 by Bill Bryans under the supervision of Michael Kelly, the Center's librarian/archivist. It comprises the personal and business records of Larry Larom plus data about the Buffalo Bill Memorial Association, conservation groups, and dude ranching in general.

Two taped interviews were made with Larry Larom. One, by Dr. Harold McCracken, is also in the archives of the Buffalo Bill Historical Center; the other, by Charles G. Roundy, is in the Charles G. Roundy Collection, American Heritage Center, University of Wyoming, Laramie. Additional information about Larom, the Valley Ranch, and the Valley School is contained in various other manuscript collections at the American Heritage Center.

Information about the Dude Ranchers Association can be located in the Larom Collection as well as in the following: 1) the Dude Ranchers Association Collection, American Heritage Center; 2) *The Dude Rancher*. Although no complete run of this magazine has been preserved, many copies may be found in several libraries, including the University of Wyoming, the Denver Public Library, The Harold McCracken Research Library of The Buffalo Bill Historical Center, and the Historical Society of Montana Library in Helena; 3) Minute Books of the Dude Ranchers Association Meetings are available for the 1929, 1930, and 1931 meetings and may be found at the Wyoming State Archives, Cheyenne, the University of Wyoming, and the Denver Public Library.

Various articles about Larom and the Valley Ranch were published in the *Cody Enterprise*. Data about the founding of the DRA in 1926 may be located in the *Bozeman Daily Chronicle*. General Information about dude ranching is in: Charles G. Roundy, "The Origins and Early Development of Dude Ranching in Wyoming," *Annals of Wyoming* Vol. 45, No. 1 (Spring, 1973).

Some information in this catalogue is taken from these books: John F. Reiger, *American Sportsmen and the Origins of Conservation* (N.Y.: Winchester Press, 1975); Harold McCracken, Dick Frost, Leo Platteter, and Don Hedgpeth, *The West of Buffalo Bill* (N.Y.: Harry N. Abrams, Inc., 1974).

The quotation at the beginning is taken from a letter, I.H. Larom to Harold McCracken, November 10, 1957, in the Larom Collection. The quotation on page 18 is from the Minutes of the Sixth Annual Convention of the Dude Ranchers Association [Minute Book, 1931], p. iii.

Some of the material in this catalogue is drawn from *Tourist Bonanza: Dude Ranching and the Development of the West*, copyright 1980, 1981 by Lawrence R. Borne. An annotated manuscript of this essay is on deposit in the archives of the Buffalo Bill Historical Center.

COLOR PLATES

M.D. Dolph, *Valley Ranch*, Oil on canvas. 8½ x 12 inches, c. 1950. Mrs. Irma Larom.

Valley Ranch School memorabilia: *The Round-up*, 1931; loving cups for team sports; Larry Larom, c. 1915; textbooks; hunting hat; and Valley Ranch flag. (Photo by Mark Payler).

Larry Larom, c. 1920.

Blackfoot buckskin shirt, decorated with beadwork and ermine tails, c. 1900. (Photo by Mark Payler).

Collection of Southwest materials: Navajo rug, c. 1940; Apache moccasins, c. 1890; Mexican style spurs, c. 1900; and Souhwestern Indian leather belt, with German silver conchos, c. 1890. (Photo by Mark Payler).

Crow woman's canvas saddle with beadwork, c. 1890. (Photo by Mark Payler)

Parfleche box and trunk, painted rawhide, c. 1900. (Photo by Mark Payler)

Sioux bonnet, eagle feathers, quill forehead band, c. 1890. (Photo by Mark Payler)

Beaded Sioux vest, c. 1910. (Photo by Mark Payler)

Boots and bronze spurs worn by Larry Larom. (Photo by Mark Payler)

Larry and Irma Larom at Valley Ranch.